I0488320

A SIMPLE EXPLANATION

Of

Crop Circles

Dee Gragg P.E.

Copyright © 2010 by C. D. Gragg

Published by
C. D. Gragg
2341 Apache
Alamogordo, New Mexico 88310

All rights reserved. Except for the inclusion of brief quotations in a review, no part of this book may be reproduced by any mechanical, photographic, or electronic process, or in the form of a photographic recording, nor may it be stored in a retrieval system, transmitted or otherwise be copied for public or private use without written permission of the author.

For additional information, contact the author at the above address or by telephone at (575) 437-5419 or by email at deegragg@yahoo.com.

The photograph on the front cover is:
Copyright © Brett Parrott. www.theroswellcode.com
All rights reserved.

ISBN 978-0-557-76488-4
SECOND EDITION
Printed in the United States of America by
Lulu Enterprises, Inc

Acknowledgements

I would especially like to thank **Mr. Jeffrey Wilson,** Director ICCRA – Independent Crop Circle Researchers' Association [International] for his expertise in reviewing and suggesting changes to this manuscript.

As always I give a special appreciation to my son **David Gragg** for his editing insight. His ability to review, revise, expand and compress, and add and subtract has always been phenomenal. Particularly valuable has been his ability to take what I said and turn it into what I meant to say.

Mr. Ted Robertson has encouraged me from the beginning with my crop circle research. Without his hard work in measuring and drawing the crop circle formations I could not have done any musical note research. Thanks Ted!

Finally, a great big hug to my wife **Nancy Sue** for proof-reading everything. She has always been my first line of defense against not only typos and spellos but my occasional nonsequential thinking and looney views. The reason you will not read those is because she took a club and beat them to death before they got out of my computer.

Table of Contents

Introduction

Crop circles have been around for a long time, perhaps for a very long time. The first serious modern day study of them began in England about 30 years ago and is still in its infancy.

When they were first studied there were two common beliefs. People either believed that all crop circles were man made or that none of them were. Unfortunately, the researchers had hardly begun before they were plagued by the hoaxers Doug Bower and Dave Chorley.

Doug and Dave offered only their word and a few crudely constructed circles as evidence to claim that they had made **all** the circles. The media didn't challenge the claim. In fact they naively published every word of it. They had a powerful story about "the men who conned the world". When it turned out that Doug and Dave were hoaxers and couldn't possibly have created more than a few circles, the media went limp and didn't bother to correct their previous, outrageous reports.

For awhile the view that "all circles are hoaxed" prevailed. But gradually the evidence forced many to change their views to a "mix of authentic and hoaxed". The prevailing question became, "how do you tell them apart"? Some are incontrovertibly authentic and some are so obviously fake that even a journalist could figure it out.

However, there in lies a great controversy which exists to this day. There were 800 crop circles discovered in 2006 alone. Are these 80% hoaxed or are they 80% authentic? Or is there some other percentage in between? The bigger question is, "How do you know"?

Historical

Crop circles have been around for a long time—at least several hundred years. They may have been around much longer than that. One of the earliest references may turn out to be from Lyon, France in 815 A.D.[1] where they were claimed as the work of the Devil. Note: the superscripted numbers refer to the references at the end of the text.

The first recorded instance of the formation of a crop circle is given in an English woodcut pamphlet[2]. See Figure 1.

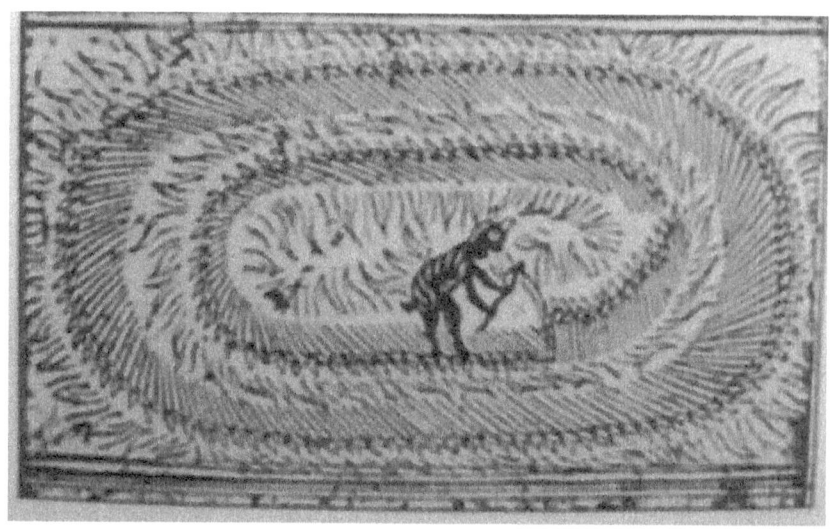

Figure 1. Mowing Devil

This pamphlet was titled "The Mowing-Devil: or, Strange News out of Hartford-Shire." Here in the illustration, the devil, complete with horns and tail can be seen creating the crop circle. This pamphlet was published in 1678.

Early Crop Circles

According to the farmers of southern England, crop circles have existed in their fields for generations. No one knew how or when they were created, they just appeared every summer. Sometimes they were laughingly called

UFO nests, but usually they were just ignored as something that happens in the fields in summer.

The early crop circles were often only a single circle or a small group of circles. They appeared in cereal grain crops such as wheat, oats, barley, and rye. The fields where they appeared had tram lines the width of tractor wheels. One could easily walk in those lines without leaving foot prints.

According to the stories, they were made in the four to five hours of total darkness which is typical of southern England summers. Since no one saw them being made it is open to speculation as to how and when they were made. One characteristic of these circles is that, the grain inside the circle was bent at 90 degrees without breaking the stalks.

While it seems that most of the crop circles were discovered in southern England, they were actually a worldwide event. Some were discovered in the United States, Germany, and France as well as other countries. Apparently no one had made any serious study of them since the late 1800s[3].

Crop Circle Research

The crop circle phenomenon as we know it did not begin in earnest until the 1970s[4]. It took another 10 years before it attracted its first serious researcher, Dr. Terence Meaden[5].

Dr. Meaden was a qualified meteorologist and the head of the Tornado and Storm Research Organization in Great Britain. He was joined in his research by Pat Delgado, a retired electro-mechanical engineer; Colin Andrews, an electrical engineer; and Busty Taylor, a light aircraft pilot.

Dr. Meaden put forth the hypothesis that crop circles were formed by what he called a "plasma vortex" which is an electrically charged spinning column of air. Dr. Meaden was widely respected and his explanation was accepted—for a while.

However, one had to strain quite a bit to accept this hypothesis because whirlwinds tended to make messy looking areas but the crop circles had cookie cutter precision walls. Further, Delgado and Andrews kept compounding the problem with the hypothesis by finding **groups** of circles. Some of the crops in the circles swirled one way and some the other. For this hypothesis to work one would now need to have several whirlwinds of different sizes and swirling in different directions, all very precisely controlled.

Dr. Meaden continued to modify his hypothesis but it was becoming increasingly difficult to fit to the facts. But, if not whirlwinds, then what was making the crop circles. The media in its usual silliness adopted the theory of little green men[5].

Sensible people put forth nonsensical ideas. Some of their ideas were: drunks with string, wild young farmers, disillusioned art students, out-of-work journalists, disinformation people from the military, over-application of fertilizer, interference from mobile phones, flocks of geometrically gifted crows or perhaps sex-mad hedgehogs[5].

Some even suggested a hot air balloon which could hover over the crop so the perpetrators could do their work without leaving any tracks. But after all the wild ideas were considered, we were no closer to an answer.

Uncommon characteristics demanded more than silly explanations. A close examination of the downed crops revealed some most unusual characteristics. Inside the circles the plant stems were not broken, but were bent at nearly a right angle just below the first node or joint of the plant stem. Even the canola plants whose stems are as brittle as celery were not broken, but looked rather steamed in place[5].

These plants did not die as the broken and crushed plants in hoaxed circles did. (We will discuss hoaxed circles in detail in a later section.) The plants continued to grow and grew with an even greater vigor[6].

The numbers of crop circles continued to proliferate. In 1990 in Wilshire alone there were 500 circles and in southern England there were a breathtaking 800[5]. They kept on coming in ever increasing complexity and numbers. They seemed to cluster around sacred sites such as Stonehenge, Avebury, Silbury Hill and others[6].

Note that at that time every circle was counted so it is actually unknown how many formations there were. Certainly many fewer than these numbers.

Then as if to completely kill the whirlwind hypothesis, pictograms began to appear in the fields. Pictograms are circles which are connected with lines and sometimes have semicircles above and/or below the circles. Some contained appendages resembling keys. They continued in number and complexity and size.

Two schools of thought developed on the origin of the crop circles. One said they were made by a higher intelligence, the other said they were all man made hoaxes. Undeterred, the crop circle researchers continued gathering data and marveling at the beauty and magnificence of the creations.

Doug and Dave
"The Men who Conned the World"

Then with the surprise of a wild bull at a garden party and the finesse of a freight train crashing through a glass factory, Doug and Dave burst on the scene.

Doug Bower and Dave Chorley were two painters who were also drinking buddies. One night in 1978 after a drinking bout, they decided they would have a little fun. They made some crop circles or as they called them "UFO nests".

They continued making a number of crop circles for the next 13 years. At that time they decided to go public with their circle making. The reason(s) why they came out remain murky to this day. Was it for fame and fortune or, at 62 and 67 were they just getting too old to stay up until the wee hours of the morning?

Others believe they were pawns of the British Government. They believe this was the government's way to ridicule any one who was doing crop circle research and at the same time take it out of the public eye. (This would be somewhat like what President Truman did to the UFO believers in 1947[7].)

So for whatever the reason, in September 1991 the British newspaper *Today* proclaimed that Doug and Dave had made all the circles since 1978! *Today* proclaimed them THE MEN WHO CONNED THE WORLD.

Doug and Dave claimed to have made the circles with a board and a rope. They said they constructed the round circles using a piece of string and made the lines straight using a piece of wire dangling from Doug's baseball cap. They made a crude demonstration circle as their proof to the media!?

The media loved it. With all the joy of a child accepting candy on Halloween, the media accepted everything these two claimed. Without bothering to check the facts, without engaging their brains as to whether these two could have actually made the hundreds of often complex circles, the media ran with the story. Both television and newspapers ran the story worldwide.

In the United States Peter Jennings gave the story to his millions of ABC TV viewers without a word as to its complete lack of any evidence and its total implausibility. (Yes, this is the same Peter Jennings who in 2005 did the dreadful disinformation special, "UFOs: Seeing is Believing"[8].)

The crop circle researchers were devastated. Some were so discouraged that they left the field. However, Colin Andrews confronted these wretches with pictorial evidence that 298 crop circles had been made from 1972 to 1980[4]. Had they made them? Well, no.

Had they ever been active around the Avebury area? No, they had never been there. Avebury had been the most active crop circle area since 1988.

When confronted with some of the more complex circle patterns Doug and Dave were not even able to draw them on paper. It was clear they had no understanding of the math involved.

How did they construct the straight lines? Did they not realize that the wire dangling from the bill of Doug's cap would not of its self let them sight a straight line. To do that would obviously take a second sight? How had they been able to continuously swirl the grain from center to circumference. This could certainly not be done with only a board and a piece of rope.

Finally, how had they consistently avoided detection by farmers, campers, researchers, guard dogs, night cameras, inferred detectors and alarm systems[5]? They still had no answers.

The real con was of the media! The media completely ignored this sorry confession and refused to tell their readers and listeners how they had reported complete nonsense to them as if it were the truth? If fact, none of this ever reached the public. To this day I have personally had people point out to me how crop circles are created by stomping the grain with a board. **The con lives on.**

Copy Cats

After Doug and Dave's retirement from the scene in 1991, crop circles continued! In fact, the number and complexity of designs continued to grow to the extent that one wondered what if anything Doug and Dave had been contributing. Whatever circles they may have been creating were never missed.

However, now there were other problem people. There had always been people who claimed that all crop circles were hoaxed. They now came out in earnest. Some sought to show how the circles were faked.

The Wessex Skeptics wanting to protect their established scientific view that everything can be explained in terms of orthodoxy, decided to make a circle one night. They had hardly begun when they were caught red-handed at Cheesefoot Head by crop watchers[2].

Not having previously obtained the farmers permission, they could have been charged with trespassing and destruction of property among other things. An anonymous donation to the farmer kept them out of trouble, but they learned quickly that hoaxing was not so easy in the middle of the night, and on someone else's property.

If Doug and Dave were not making the circles, who was? It turns out that there was no shortage of groups who claimed to have taken up the work. In fact just as Doug and Dave did, many also claimed to have **made 'em all**.

For various reasons they didn't want to claim which specific formations each had made or let anyone observe them being created. Some of their reasons were: they were secret artists, or they didn't want to be arrested for trespassing, or they didn't want to be sued for criminal damage to crops, etc. but you get the idea.

Let us examine one of their claims. Across from Stonehenge one Sunday afternoon in 1996 in broad daylight a formation of 151 circles appeared. The formation measured 915 feet from top to tail. This famous formation has been given a special name, the "Julia Set".

We know from witnesses that it appeared in a 45 minute window from the last time that someone saw the field without it and the time it was discovered[4]. So, how did they fake this in broad daylight? You be the judge. It took **a team of eleven surveyors five hours just to measure the design!!** [10]. We will look at some more of their claims in the next section.

It would be nice if we could now simply dismiss the copy cats and move on with the research. But the pesky little rascals will not go away. In fact, they have continually improved their fakes and now often look similar to the authentic circles. We will address that in the following sections.

Authentic Circles

After Doug and Dave there has been continuous disagreement as to the authenticity of many of the circles. Were they hoaxed for fun? Were they hoaxed by the British government or the media to use as a "gotcha"? Both did that at times. The real question is "Which ones are authentic"? And how do you know? We will use this section to address those which we know to be authentic.

Eyewitnesses

There have been more than 50 eyewitnesses to crop circle formations. Only one of those has been in the United States. We will use that one as our first example.

On the morning of July 4, 2003 Mr. Art Rantala was having his first cup of coffee and watching a weather front move into the area[11]. At about 7:40 A.M. with the rain falling and the wind beginning to pick up he noticed a group of trees across the street started swinging every which-way. Then he saw the crop circles begin to form in the wheat across the street.

The circles formed one at a time, the right one first, then the left one and finally the one in the middle. They appeared as "black-holes" when they formed. All of the circles were made in less than 15 seconds. Whatever it was that caused the circles to form, it was beyond the range of visible sight.

Figure 2 shows the circles that were formed. In my analysis of this formation I noted that it had an "inscribed" (Points A , B, C) and "circumscribed" (Points D, E, F) equilateral triangle which is Dr. Hawkins Theorem II. Note, that I **only used** Dr Hawkins Theorem II. I made no attempt to prove it. That came later in my metamorphic development and is included in the Musical Notes in the Fields Section.

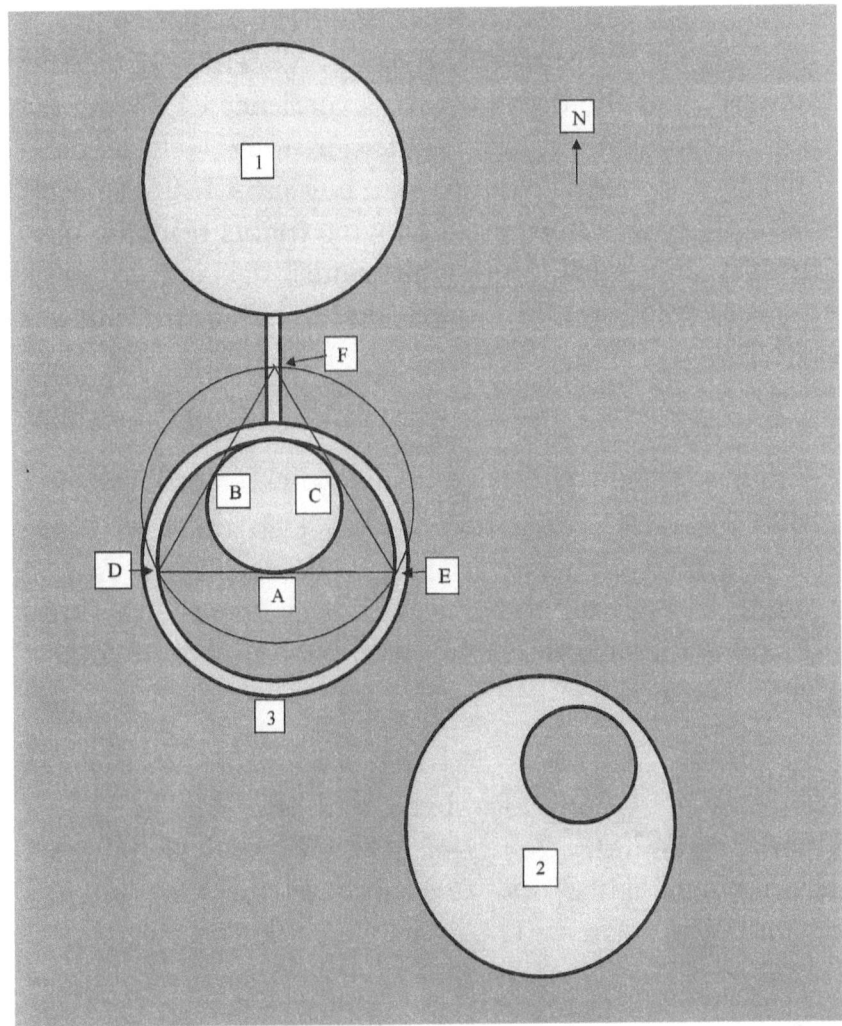

Figure 2. Kekoskee/Mayville

(On a personal note, this was my first time to work with Mr. Jeff Wilson and the rest of the Independent Crop Circle Researchers' Association [International]. This also was the first time I tried to apply Dr. Gerald Hawkins work to crop circles.)

One couple in Surrey, England, Vivien and Gary Tomlinson, had the disconcerting experience of having a small pictogram appear around them in 1989. They were thrown by a strong force into standing crop from a near by path. They described a tall glowing funnel reaching up to the sky and small "mini-whirlwinds" dancing around their feet, spinning the stems down at an incredible speed[4].

Nancy Talbott has listed six eyewitness accounts on the BLT Research Team. Inc. Site[12]. Two are from The Netherlands, two from Poland, and one each from England and France. Nancy was an eyewitness to one of these and is the only researcher to have seen a crop circle formation created.

Nancy was visiting Robbert v/d Broeke in The Netherlands in 2001. At about 3:00 A.M. both saw three brilliant white lights, one at a time, as the crop circle formation was created. The lights were 100 feet away in a string bean field behind Robbert's house, yet they lit the inside of his house as if it were day[12].

Robbert is a very special person and Nancy tries to visit him for three weeks each summer[13]. Robbert has had some communication with extraterrestrials who have told him that they create the crop circles. They often inform him when and where the circles will be created.

Robbert has taken several photos of the extraterrestrials. They appear on film as smoky, ghostly images. In contrast to the crisp, clear image of the little grey in Ryan Wood's book MAJIC Eyes Only[14]. Robbert has thus far only been able to convince one of them to appear although there are a number of others. These extraterrestrials sometimes appear on film when they can't be seen with the eye.

Robbert has published a book about his experiences which has not yet been translated into English. Nancy says that as soon as it is translated she will alert us on her web site[12].

Non Cereal Circles

A number of circles have appeared in non cereal grain crops. For example circles have been reported in carrots, potatoes, string beans, tobacco, and spinach to name a few. In general these crops do not have tram lines to hide foot prints, but still they might be subjected to the same hoaxing as the cereal grain crops, just with more difficulty.

There are several mediums however, that defy even the most dedicated hoaxers. For example a number of circles have been reported in snow and ice.

On the second of February 1993 students at MIT in Boston reported snow circles[6]. They were formed in light snow that covered a very thin layer of ice on the Charles River. The ice would not have supported a small child, plus there were no footprints in the snow.

There have also been a few cases of circles in trees. At Grand Lake, New Brunswick, Canada circa 1966 there appeared two tree circles[6]. Each was about 12 feet in diameter.

The circles were in pine trees which were about three inches in diameter at the base. The trees were bent at the ground and were laid all the way over without breaking. Oddly, all the trees pointed to the circumferences of the circles.

BLT (Not a sandwich)
Expulsion Cavities and Enlarged Nodes

Let's get this straight from the beginning. BLT does not stand for bacon, lettuce and tomato. It stands for Burke, Levengood, and Talbott. They were John Burke, Wm. C.

Levengood and Nancy Talbott of the original BLT Research Team, Inc[12]. Burke and Levengood have now both left BLT. Ms. Talbott remains the president. Yes, this is the same Nancy Talbott mentioned earlier as a friend and frequent visitor of Robbert v/d Broeke.

The BLT Research Team conducts research on the physical changes on the crop circle plants. They conduct a number of tests. The two most important for quick identification in the field are expulsion cavities and enlarged nodes. Expulsion cavities and enlarged nodes **are found only inside the crop circle.** They are not present outside of it.

Expulsion cavities are holes blown out at the plant stem nodes. These nodes are the fibrous knuckle-like protuberances found along the plant stem. The cavities are usually found in the 2nd node beneath the seed-head, but may be in the 3rd and 4th nodes. These cavities could be caused by an internal heating similar to that caused by microwaves.

The plant stem nodes are often both laterally enlarged ("fatter") and longitudinally enlarged ("stretched"). The node elongation is a permanent effect related to the formation energies, and this is the parameter now used to authenticate crop circles. Again, the expulsion cavities and enlarged nodes are found only inside the crop circle and are not present outside of it.

Dr. Charles Lietzau has taken the node elongation a step further[15]. Dr Lietzau has invented a process he has named L-NEAT which stands for Levengood Node Elongation Analysis Test. This statistically-significant node-elongation is the conclusive fingerprint that represents the only scientifically definitive litmus test for an authentic crop formation. The test relies upon the Student's t Distribution.

In my statistics book the Student's t Distribution is in Chapter 9 on Page 224[16]. It has been years since my statistics class so it took me about an hour to relearn the Student's t Distribution and another hour to apply it to some typically expected data. This will not be for everyone.

Dr. Eltjo Haselhoff has also contributed to our further understanding of node elongation by studying its distribution inside the circle[17]. Beginning at the edge of the circle the node elongation increases to a maximum at the center of the circle. Continuing in a straight line (the circle diameter) the nodes decrease as they get nearer the edge of the circle.

Time Constraints and Complexity

The summer nights in England are short and have only four to five hours of complete darkness. This is when most of the crop circle formations appear. The hoaxers typically claim that this is plenty of time.

However, a few of the most complex and beautiful formations are constructed in daylight hours and in fields where occasionally people are present. Typically the people don't see them being constructed but discover them soon afterwards.

We have already considered the Julia Set. It was constructed across from the popular tourist attraction Stonehenge on a Sunday afternoon in 1996. This formation of 151 circles was constructed in broad daylight in 45 minutes. The formation measured 915 feet from top to tail and took a team of eleven surveyors five hours just to measure the design.

. A few weeks later at Windmill Hill near Avebury "The Triple Julia" formation appeared[9]. It was created in one night with a maximum darkness time of five hours.

It contained 194 circles and was essentially three Julias with many of the small inner circles replaced by one large circle. It was 858 feet by 858 feet. **A surveying company who analyzed the site quoted a minimum of five days just to mark the site alone**[10].

Finally, the crop circle formation to top them all. The formation was perched atop the downs of Milk Hill near Alton Barnes[4]. It was the largest single diameter and most elaborate glyph ever to have appeared. It has the appearance of the Julia formations except that it is a six-armed design. It is called the Katherine Wheel.

It had 409 circles and was around 800 feet wide. The circles ranged in size from 70 feet to 4 feet in diameter. It came during a night of rain, yet was found pristine the next morning, without mud splatters or soil disturbance.

Significantly none of the hoaxer groups stepped forward to claim to have made this formation. With an air of humility, one member of the usual human public circle making crews even issued a statement expressing severe doubts as to how such a design could be laid down in one night.

Probably Authentic Circles

Many circles are indeed truly authentic. Many are clearly hoaxed. It is that middle area where there is much disagreement. Many circles just do not fall into a nice positive identification. However, there are characteristics that we expect to find in authentic circles. But be careful, those characteristics by themselves are generally not sufficient to verify a circle as authentic.

There are no tracks through the crop up to the circle. Of course, tracks may be caused by tourists or others who have been to the crop before you. Also in English crops there are tram lines in which the hoaxers can walk and often leave no prints.

There are no signs of interference with the soil or crop where hoaxers could have stood to make the formation. There should also be no pole or stake marks that would have been necessary for a hoaxer to lay out the formation.

The plants are bent over without breaking or crushing. Be aware that some of the most experienced hoaxers have learned how to do this in all crops except canola, maize and soybeans.

An authentic circle will likely be symmetrical. If it has a number of arms or circles or partial circles they should be in exact relationship to each other. The walls of the circles are often so well defined that they have a cookie cutter preciseness. The crop is swirled symmetrically with two or fewer swirls to the circle circumference. Again the hoaxers are making better counterfeits, so that none of these characteristics would alone make the formation authentic or their absence make it a hoax.

Magnetic anomalies can be conclusively recorded. Colin Andrews found in his hoaxing study that 54 of the 236 crop circles had magnetic anomalies.

Animals, BOLs, and Electronics

Animals

There is an entire smorgasbord of events which sometimes occurs before, during or after an authentic circle formation. None of these events of themselves are proof of an authentic circle. Also, their absence is not proof of a hoax.

Farm animals and pets often behave in unusual ways. Dogs bark incessantly and refuse to stop[5]. Dogs are particularly sensitive to ultrasound, as evidenced by their reaction to pre-earthquake conditions. When dogs are particularly anxious, they mark their territory. They will find the energy centers of crop circles and treat them as they would a fire hydrant.

Cats also react strangely. Normal placid cats suddenly become agitated when they enter a circle. And just as soon as they leave the circle they regain their composure.

Cattle and sheep act in a distressed manner prior to the arrival of a circle. Sheep will attempt to move as far away as possible from a particular field before a formation appears. Horses refuse to cross the perimeters of crop circles, or they become nervous in their vicinity.

Dead animals are rarely found inside crop circles where you might expect, for example field mice[4]. There are a few exceptions: Dogs have been found dead in two instances in the USA. In Canada two dead, flattened porcupines were discovered. Porcupines tend to roll in a ball and stand their ground. However skunks and turtles also stand fast and they have not been found in a crop circle.

Opossums tend to faint if threatened and they have not been found either. Perhaps even these critters have some early warning. Linda Moulton Howe did report some flies have been found dead in a formation with their tongues stuck to the wheat[19].

Balls of Light (BOLs)

Balls of Light (BOLs) are basketball size lights often orange, yellow or white in color. They have been flying around since at least the Second World War[7]. They often flew with American pilots where they were called "Foo Fighters". They also flew with the German and Japanese Air Forces.

BOLs are often associated with crop circle formations[5,6,17,19]. They have been seen before, during and after the formation of a circle. There have even been implications that they have created the circles.

They may have a more mundane purpose. It has been theorized that the BOLs are simply information gathering devices for a stand-off UFO.

Electronics

Often cameras and camcorders may work perfectly well outside a crop circle formation and go completely dead inside[4,19]. This seems to be due to a power drain from the batteries. Occasionally, inexplicably, when they are taken outside the circle, the batteries power up and the cameras and camcorders work perfectly. On occasion they never work again.

Mr. Jeff Wilson, Director of the ICCRA has lost two GPS devices in crop circle formations. In both cases, the GPS screen began to produce a faint electrical glow before the electronics were burned out.

Mobile telephones also lose power but even if power is restored they may not function due to apparent interference with their transmission frequencies[4]. Watches have been known to drastically slow down or gain time[4,17].

Magnetic compasses sometimes point in the wrong direction or just spin wildly. Crop harvesters have been known to breakdown when crossing a circle formation[4]. Mysterious sounds have been heard and recorded in the formations[4,6].

The "80/20" Statement

Colin Andrews

Colin Andrews states in his book ***Crop Circles, Signs of Contact*** that in the 1990s he became aware of a growing number of manmade crop circle formations[6]. In 1999 he began an investigation into hoaxing.

The project was fully funded by philanthropist Laurence Rockefeller and took place during 1999 and 2000. His research included engaging retired police detectives as private investigators; crop circle site inspections; collection of physical evidence; extensive aerial photography; recording of personal experiences; and gathering information from the media.

From his research he concluded that approximately 80 percent of all the crop circles investigated in England from 1999 through the year 2000 were manmade. On

August 9, 2000 he announced his findings live on BBC National News in London.

This announcement changed his life. Since making that announcement, he has received hate mail, threats, red-faced zealots screaming in his face, contempt, media ridicule, professional scorn, hatred, and insults. People who believed that **all** crop circles were of paranormal origin thought their world had come to an end.

The furious backlash heaped upon him made their despair quite clear. He was accused of being a funded disinformer in the employ of the Rockefeller family or that he was being paid by the CIA to deliberately spread disinformation.

However, Mr. Andrews has a major point to make. He found that 20 percent of crop circles are real, unexplained formations and should be considered a gift to mankind. To my knowledge Mr. Andrews has never backed off of his original 80/20 statement nor have his detractors tired of insisting that the statement is wildly inaccurate.

Hoaxed Circles

In a previous section we have seen how to identify authentic circles. We have also noted those characteristics that indicate that a circle is probably authentic. However, there is often great difficulty in determining which circles are authentic and which are hoaxes.

Even competent researchers may disagree over this. Surely no one could hope to make a good judgment of this without being at the formation. However, just as there are characteristics of certain authentic circle formations, there are characteristics of certain hoaxed formations.

The first thing to look for is footprints. Footprints from the edge of the field into the crop circle formation are a tell-tale sign of a hoax. In England, hoaxers use tram lines to avoid footprints but in the US, many times they don't have that help. In US circles that is a sure hoaxing give away. One must be aware to be sure you are the first one into the circle or that you know the footprints are those of previous viewers.

The edges of the circles should have cookie cutter precision. The edges must not have a "serrated edge".

Mr. Jeff Wilson, Director of the ICCRA, says serrated edges are produced when the crop is pushed straight forward by a plank or garden roller rather than in a circle[20]. Then the correction produces the serration at the edge of the circle. Further, if the crop in the circle is not swirled but is downed in concentric circles this is the mark of a plank or roller.

The next thing to look for is holes from a pole or stake used as the center of each circle or portion of a circle. The hole may be covered or camouflaged but it will be clear where it must be.

Finally the crop must not be crushed or broken in the flattening. Drawing from the experience of Mr. Wilson[21] , when inspecting a hoaxed formation, he wrote, "About 95-98% of the soybean stalks were broken or snapped;"

Musical Notes in the Fields
The Work of Hawkins and Gragg

Dr. Gerald Hawkins is best known for his work with Stonehenge and most particularly for his astounding book **Stonehenge Decoded**. However, those of us who do crop circle research are most indebted to him for his work in decoding and proving Euclidean geometric theorems and for his work in decoding musical notes from the crop circles.

For his research Dr. Hawkins used the data from Mr. Colin Andrews pioneering best-seller **Circular Evidence**[22]. These data were the only set of accurately measured circle diameters available for rigorous analysis for the early period of the phenomenon[6,23].

Dr. Hawkins discovered four new geometric theorems which were decoded from the crop circles. (There is a fifth theorem which I will deal with later.) Amazingly, none of these theorems appeared in Euclid's Elements[24]. These were **new theorems which were unknown to the world of mathematics** until that time!

Equally amazing was that **all four of the theorems contained diatonic ratios!** Pythagoras developed the octave in music. Its development relied upon the ratios, of the notes to middle C. These are called diatonic ratios (Figure 3). For example the next higher note is D which has a diatonic ratio of 9/8.

Note Name	C	D	E	F	G	A	B	C
Diatonic Ratio	1	9/8	5/4	4/3	3/2	5/3	15/8	2
Frequency (Hz)	264	297	330	352	396	440	495	528

Figure 3. Diatonic Ratios

It is 9/8 higher in frequency at 297 Hz. E is 5/4 at 330 Hz etc. This is the procedure for diatonic ratios in all octaves both higher and lower in frequency. The notes which Dr. Hawkins discovered are shown in Figure 4 in red. Note that Theorem IA and Theorem IVA discovered the same note, F at 352 Hz.

Note Name	C*	D	E	F	G	A	B	C
Diatonic Ratio	1	9/8	5/4	4/3	3/2	5/3	15/8	2
Frequency (Hz)	264	297	330	352	396	440	495	528

Note Name	C	D	E	F	G	A	B	C
Diatonic Ratio	2	9/4	5/2	8/3	3	10/3	15/4	4
Frequency (Hz)	528	594	660	704	792	880	990	1056

Figure 4. Euclidean Diatonic Ratios

Next Dr. Hawkins looked for diatonic ratios and their notes and frequencies among the circle diameters. He took the diameters of the large circles and divided them by the diameter of the satellite (smallest) circle and used this result to determine any integers.

Using the integers found, he searched for diatonic ratios by raising 2 to the n/12 power where n is the integer just found. For example if:

Large Circle = 10   Satellite Circle = 2

Then: $10/2 = 5$; And the diatonic ratio = $2^{5/12} = 2^{.4166} = 1.335$

This is the diatonic ratio 4/3, which is F just above middle C.

Using this procedure Dr. Hawkins was able to find the entire octave above middle C, Amazingly, he found only diatonic ratios (the white piano keys) and found no nondiatonic ratios (the black keys). All of Dr. Hawkins diatonic discoveries are shown in Figure 5.

Note Name	C*	D	E	F	G	A	B	C
Diatonic Ratio	1	9/8	5/4	4/3	3/2	5/3	15/8	2
Frequency (Hz)	264	297	330	352	396	440	495	528

Note Name	C	D	E	F	G	A	B	C
Diatonic Ratio	2	9/4	5/2	8/3	3	10/3	15/4	4
Frequency (Hz)	528	594	660	704	792	880	990	1056

Figure 5. Hawkins Diatonic Ratios

Dr. Hawkins died May 26th 2003 at the age of 75. He has been sadly missed by all in the crop circle community.

This was just about the time I became interested in his work.   I have some background in math (mechanical engineer) and I wanted to see the proofs of his theorems. That seemed like a reasonable request-- until I started to look.

I went to a quite prominent author to look for the proofs. He had the figure for each theorem and the hypothesis stated as if it were a conclusion. He showed no proof whatever of the theorem.

Now every theorem must have a hypothesis, followed by the steps of its proof. If the proof steps are correct the conclusion will now be the same as the hypothesis, thus proving the theorem.

I went to another author who likewise provided no proofs. He treated the hypothesis as if it was the conclusion and wrote blissfully on. I searched more authors, some prominent and some not so prominent. But the pattern was always the same.

It was maddening. All the authors apparently assumed that no proofs were needed. Indeed, I had my doubts that some even recognized that there were proofs. Now, I never doubted Dr. Hawkins proofs of the theorems, I just wanted to see them.

It was in this hour that I truly missed Dr. Hawkins, a man I had never met. But Dr. Hawkins was gone and there was no place to turn. So, I made a fateful decision. I had been a fair student in 1948/49 as an 11th grader in Miss Stella Edmiston's plane geometry class. **I would prove the theorems.**

As the old cliché goes, that was **much** "easier said than done". In fact this cliché became the mother of all clichés for me. This was a monstrous, lunch eating task! I worked for three months to prove those theorems. I found all the pits and fell into most of them.

For example, I had proved a theorem using the square and one using the hexagon, so I was sure there must be one for the octagon. If there is a theorem for 4 sides and one for 6 sides, there must be one for 8 sides. That's only common sense. Right? **Wrong!**

I worried over that one for three weeks. No matter what I tried I always had one more unknown than there were equations—rendering it always unsolvable. Finally in desperation, I invented what I call Theorem T (Appendix A). This now shows conclusively that the only regular sided polygons which are diatonic are the equilateral triangle, the square, and the hexagon.

This now brings me to Dr. Hawkins Fifth Theorem. It stated that all the diatonic theorems could be derived from it. Of course, but the only ones possible; the equilateral triangle, the square, and the hexagon, he had already derived. The Fifth Theorem was esthetically beautiful but not very useful.

I must admit this had been fun and along the way I discovered five more theorems. These theorems were also all previously **unknown to the mathematical world.** I wrote all nine of them into a technical report[25].

I had published a number of other research papers while employed by the Air Force. But, I had no idea what the reaction of the crop circle community would be to this. So, finally I just took a deep breath and sent it out.

The reaction turned out to be overwhelmingly favorable. In fact, to this day no one has contacted me to point out an error in any of the proofs! I hope that doesn't mean they just weren't reading them. Indeed, from this publication came one of my most prized pieces of paper:

An Email from Mr. Colin Andrews which read in part[26]:

"Gerald would be very pleased to know that you are taking up this particular work from where he had left off."

I thought **"wow"** I hadn't planned on taking up any one's work, I was just having fun. But then, if the most prominent crop circle researcher in the world thinks I can do it, maybe I can. Maybe I should. So I decided I would give it a go.

Dr. Hawkins had done all of his research on crop circles in England. I had no access to data from England. However, thanks to Mr. Jeffery Wilson and Mr. Ted

Robertson I did have access to data from the USA. How would the data from the two compare?

For my first research Ted sent me the data from the Locust Grove, Ohio, crop circle formation. This was a most fruitful formation. It contained four diatonic ratios: 5/4 which is E at 330 Hz; 3/2 which is G at 396 Hz; 9/4 which is D at 594 Hz; and 8/3 which is F at 704 Hz.

The E and G had previously been discovered by Dr. Hawkins in the English crops. So, now I had duplicated his work here in the USA. Additionally, there were two previously undiscovered notes D and F. The formation contained only diatonic ratios (the white keys) and no nondiatonic ratios (the black keys). This was a tremendous confirmation here in the USA of his work in England[27]

.

Next I analyzed the data from the Miamisburg, Ohio, formation[28]. It contained four notes previously seen by Dr. Hawkins. But then it contained a most startling surprise. **It contained a nondiatonic ratio!?** This was the first ever in 23 years. Specifically, it contained 19/12 which is G# or Ab at 418 Hz.

That wasn't the end of it. In the Geneseo, Illinois, formation there was another nondiatonic ratio 17/16 which is C# or Db at 281 Hz[29]. The remaining formation, Madisonville contained no more surprises[30].

A summary of all notes discovered to date are in the following Figure 6.

Diatonic Frequencies in the Fields

Note Name	C	D	E	F	G	A	B	C
Diatonic Ratio	1/4	9/32	5/16	1/3	3/8	5/12	15/32	1/2
Frequency (Hz)	66	74.25	82.5	88	99	110	123.75	132

Note Name	C	D	E	F	G	A	B	C*
Diatonic Ratio	1/2	9/16	5/8	2/3	3/4	5/6	15/16	1
Frequency (Hz)	132	148.5	165	176	198	220	247.5	264

Note Name	C*	D	E	F	G	A	B	C
Diatonic Ratio	1	9/8	5/4	4/3	3/2	5/3	15/8	2
Frequency (Hz)	264	297	330	352	396	440	495	528

Note Name	C	D	E	F	G	A	B	C
Diatonic Ratio	2	9/4	5/2	8/3	3	10/3	15/4	4
Frequency (Hz)	528	594	660	704	792	880	990	1056

Nondiatonic Frequencies in the Fields

Note Name	C#/Db	D#/Eb	F#/Gb	G#/Ab	A#/Bb
Nondiatonic ratio	17/16	19/16	17/12	19/12	85/48
Frequency (Hz)	281	314	374	418	468

Figure 6. All Diatonic and Nondiatonic Ratios to Date

UFOs

Crop circle researchers have long reported that UFOs have been seen at crop circle sites. Usually the UFOs have been the little balls of light (BOLs). When viewed the BOLs usually seem to be inspecting the beautiful work in the crops.

However on at least one occasion a UFOs has been seen creating a crop circle formation[12]. In 2006 in the Netherlands Robbert van den Broeke and Nancy Talbott observed a "black cloud" low over a field, then a black spiral went down into the ground. A new crop circle in the form of a "Celtic Cross" was found immediately afterwards.

There are several instances in which a UFO has been sighted in a field and the next day a crop circle formation was found. For example, in Poland in 2000 farmer Jerzy

Szpulecki observed a red sphere descend into a wheat field. The next day a circle was observed[12].

In another example, bicyclist Mike Booth in England in 2005 observed dome shaped objects in a crop field. The next day a geometric crop circle was found[12].

On at least two occasions extraterrestrials have told us they are creating the crop circle formations. As previously mentioned, Nancy Talbott's friend Robbert v/d Broeke communicates with the extraterrestrials[13].

The ETs have told Robert that they create the crop circles. His verification of this is that they often inform him when and where the circles will be created.

In another instance, Alton Kamadon was taken aboard a space craft. He was told by the extraterrestrials that they made the crop circles[31].

First, let's take a look at the capabilities of UFOs[7]. Could they be creating the crop circle formations? They can fly at speeds of at least 6,000 mph. They can turn at 90 degree angles which means they have obviously mastered anti-gravity or they would destroy both themselves and their craft.

They can fly straight up, hover, disappear and reappear. They have been reported at lengths of more than a half mile. The most important characteristics of a UFO for crop circle making are its ability to hover and to remain invisible.

Years ago someone suggested that crop circles were made from hot air balloons because the formations were created without leaving tracks. Well, of course, that was just silly. But a UFO **really could** hover and stay in place for as long as needed.

The UFOs ability to disappear is legendary as the pilots who have chased them will testify. How do they disappear? Of course we don't know but one simple way would be to change their electromagnetic frequency.

The visible light spectrum is very narrow. All they would need to do to disappear and reappear is simply change their frequency slightly up or down. Obviously, any intelligence which has all these capabilities would have no difficulty in creating the complex and beautiful crop formations found in the fields.

Epilogue

The serious study of crop circles began in England and is less than 30 years old. But even in this brief period, important research has been hampered by hoaxers.

The identification and importance of the hoaxers has caused significant disagreement among even the serious researchers. Indeed, one of the most prominent researchers has confidentially advised me not use **any** of the crop circle data from England in my research.

The research in the USA began later than in England. But, through the work of the BLT Research Team it has become easier to identify the authentic crop circle formations. There are considerably fewer crop circle formations in the USA than in England. Fortunately, BLT accepts samples from all over the world and continues to separate the authentic from the hoaxes.

When will we ever know who is creating the authentic crop circles? Curiously, while the US government has taken the lead in suppressing information about UFOs, the English government has taken the lead in suppressing information about crop circles.

Both UFOs and crop circles make governments and people nervous and uneasy. Governments clearly have huge amounts of information which they will not release. People tend to limit their awareness to what they already can understand. The amount of uncertainty surrounding unexplained phenomena leads people to ignore it and "let it go away".

Both crop circles and UFOs seem inextricably linked. When we learn one, we almost certainly will learn the other. Further, if the "secrets keepers"—many, most probably, with serious conflicts of interest--continue to control the secrets, it will be a very long time before we learn them.

References

1. Meeting, Bishop of Lyon with King of France, Lyon, France, 815 A.D.

2. The Mowing-Devil, English wood cut pamphlet, "Strange News out of Hartford-Shire", August 22, 1678.

3. Capron, J. Rand, Nature, circa late 1800s.

4. Thomas, Andy, "Vital Signs, A Complete Guide to the Crop Circle Mystery and Why It Is NOT a Hoax", Frog, Ltd., Berkeley, California, 1998.

5. Silva, Freddy, "Secrets in the Fields", Hampton Roads Publishing Company, Inc., Charlottesville, VA, 2002.

6. Andrews, Colin, "Crop Circles, Signs of Contact", New Page Books, Franklin Lakes, N.J., 2003.

7. Gragg, Dee, "A Simple Explanation of UFOs", Self published, 2006.

8. Jennings, Peter, "UFOs: Seeing is Believing", ABC Television, February 24, 2005.

9. Allen, Marcus, "Behind The Hoaxers—Physicists, Scientists, Stompers and the Secret History of Circle Faking", Sussex Circular #33, September, 1994.

10. The Anatomy of Deception, "The Crop Circular" www.lovely.clara.net/circlemakers.html.

11. Wilson, Jeffrey, M.S., Charles Lietzau, Ph.D., Gary Kahlhamer, and Roger Sugden, "Initial Field Report of the July 4, 2003 Kekoskee/Mayville, Wisconsin Crop Circle Formation", www.abduct.com/features/f42.htm, February 12, 2004.

12. Talbott, Nancy, BLT Research Team, Inc., www.bltresearch.com/eyewitness.html.

13. Talbott, Nancy, "The Boy From Holland: Consciousness and Crop Circles", 15th Annual UFO Congress Convention & Film Festival, 2006.

14. Wood, Ryan S., "MAJIC Eyes Only", Wood Enterprises, 2005.

15. Lietzau, Charles, Ph.D., "Scientifically Determining the Authenticity of a Crop Formation through use of the

L-NEAT Process",
http:/psiapplications.com/Treepad/documents/128.html.

16. Mendenhall, William, "Introduction to Probability and Statistics", Duxbury Press, Belmont, California, 1971.

17. Haselhoff, Eltjo, H., Ph.D., "The Deepening Complexity of Crop Circles", Frog Ltd., Berkeley, California, 2001

18. Anderhub, Werner, and Hans Peter Roth, "Exploring the Designs & Mysteries, Crop Circles", Lark Books, New York, NY, 2002.

19. Howe, Linda Moulton, "Mysterious Lights and Crop Circles", Pioneer Printing, Cheyenne, Wyoming, 2002.

20. Jeff Wilson Email to Dee Gragg, http://us.f309.mail.yahoo.com/ym/ShowLetter?Msgld=8 336_0_99509,

July 1, 2005.

21. Jeff Wilson Email to Dee Gragg, http://us.f309.mail.yahoo.com/ym/ShowLetter?Msgld=4 698_0_18305,

August 11, 2004.

22. Andrews, Colin and Pat Delgado, "Circular Evidence", Harvard University Press, Cambridge, Massachusetts, 1969

23. Hawkins, Gerald, "Circles Phenomenon Research International Newsletter", Volume 5, No. 2, Fall/Winter, 1996/97.

24. Euclid, "Euclid's Elements", Alexandria, circa 300 B.C. The entire 13 volumes may found at the following site. http://aleph0.clarku.edu/~djoyce/java/elements/elements.html. The volumes with comments and guide sections have been provided by D. E. Joyce, Department of Math and Computer Science, Clark University, Copyright 1996, 1997, 1998.

25. Gragg, Dee, "Crop Circle Theorems, Their Proofs and Relationship to

 Musical Notes", the Crop Circle News, 2004, www.swimoutsidethepool.com, 2007.

26. Colin Andrews Email to Dee Gragg, http://us.f117.mail.yahoo.com/ym/ShowLetter?MsgId=2220_1730548, June 10, 2004.

27. Gragg, Dee, "Analysis of the Locust Grove Crop Circle Formation - A Study To Determine Its Relationship To Musical Notes", Crop Circle News, August 19, 2004. Also, The Circular, Crowes Complete

Print, Norwich, Norfolk NR6 6JB, Winter-Spring 2005, p. 67.

28. Gragg, Dee, "Analysis of the Miamisburg Crop Circle Formation - A Study to Determine Its Relationship to Musical Notes", Crop Circle News. Also, The Circular, Crowes Complete Print, Norwich, Norfolk NR6 6JB, Winter-Spring 2005, p. 73, www.swimoutsidethepool.com, 2007.

29. Gragg, Dee, "Analysis of the Geneseo Crop Circle Formation - A Study to Determine Its Relationship to Musical Notes", Hamilton House, www.alienabductionhelp.com, 2006, www.swimoutsidethepool.com, 2007.

30. Gragg, Dee, "Analysis of the Madisonville Crop Circle Formation - A Study to Determine Its Relationship to Musical Notes", Hamilton House, http://www.alienabductionhelp.com/htm/Dee_Gragg_2. htm, 2007, www.swimoutsidethepool.com, 2007.

31. Eagle Wings Magazine, Star Dreams, Genesis Communications Corporation, Vancouver, BC, 2002.

Appendix A

Theorem T

Trigonometry can be used to solve circular relationships for inscribed and circumscribed regular polygons for polygons of any number of sides from 3 to infinity.

Proof:

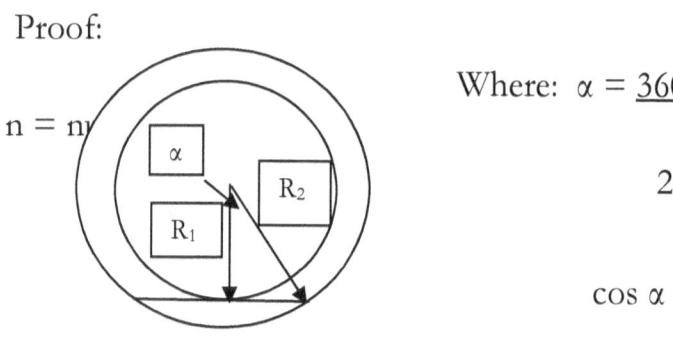

n = n

Where: $\alpha = \dfrac{360^0}{2n}$ and

$$\cos \alpha = \dfrac{R_1}{R_2}$$

$$\dfrac{R_2}{R_1} = \dfrac{1}{\cos \alpha}$$

Proving the Theorem

Figure 1. Regular polygon with any number of sides

Further: $\dfrac{(R_2)^2}{(R_1)^2} = \dfrac{(1)^2}{(\cos \alpha)^2}$

Table of Some Common Polygons

Figure (All are equiangular)	Number of Equal Sides	Ratio of Diameters and Radii	Ratio of Areas, Diameters Squared, and Radii Squared	
Triangle (1)(4)	3	2.000	4	4.000
Square (2)	4	1.414	2	2.000
Pentagon	5	1.236		1.527
Hexagon (3)	6	1.155	4/3	1.333
Heptagon	7	1.110		1.232
Octagon	8	1.082		1.172
Nonagon	9	1.064		1.132
Decagon	10	1.051		1.106
	15	1.022		1.045
	20	1.012		1.025
	50	1.002		1.004
	100	1.000		1.001
	200	1.000		1.000
	∞	1.000		1.000

(1) Theorem II, by Dr. Hawkins using Euclidian Geometry
(2) Theorem III, by Dr. Hawkins using Euclidian Geometry
(3) Theorem IV, by Dr. Hawkins using Euclidian Geometry
(4) Found in the Kekoskee/Mayville, Wisconsin Crop Circle
Formation July 9, 2003

All of my papers are on my web site www.swimoutsidethepool.com.

If you have any questions or comments please contact me at:

Dee Gragg

deegragg@yahoo.com

(575) 437-5419

A Short Bio

Mr. Gragg teaches UFOs, Crop Circle and Extraterrestrial classes at New Mexico State University, Alamogordo Community Education. He has written two short books, "A Simple Explanation of UFOs" and "A Simple Explanation of Extraterrestrials" for use in the classes.

He has published five research papers on decoding Euclidian geometric theorems and musical notes from crop circles. He has written and published a song using all of the notes and only those notes decoded from the crop circles, "The Diatonic Crop Circle Song."

Mr. Gragg often speaks on UFOs, crop circles and extraterrestrials at symposia and other events. He is a member of UFO Educators. He is a MUFON Field Investigator and is the MUFON Assistant Director for the state of New Mexico. He is an international authority on anthropomorphic dummies and escape systems and is licensed to practice as a Registered Professional Engineer.

Dee Gragg, P.E.

www.ingramcontent.com/pod-product-compliance
Lightning Source LLC
Chambersburg PA
CBHW032342200526
45163CB00018BA/2223